SCIENTIFIC AMERICAN EDUCATIONAL PUBLISHING

GRAVITY'S PULL

10 FUN GRAVITY EXPERIMENTS

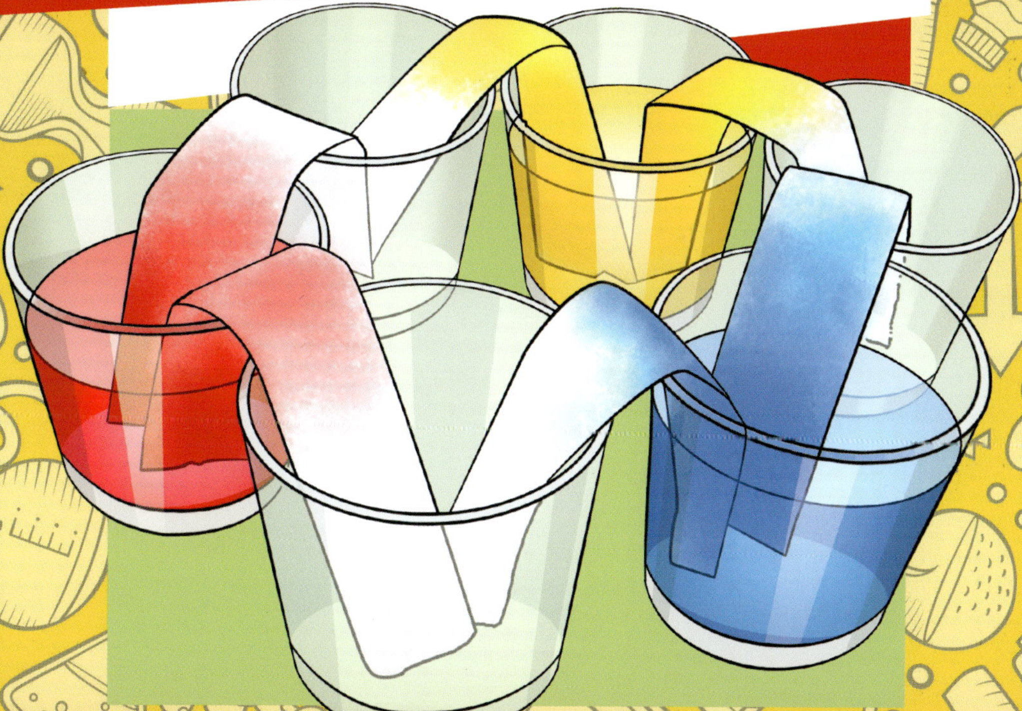

BRING SCIENCE HOME

Published in 2023 by Scientific American Educational Publishing in association with
The Rosen Publishing Group 29 East 21st Street, New York, NY 10010

Contains material from Scientific American , a division of Springer Nature America, Inc., reprinted by permission as well as original material from The Rosen Publishing Group.

First Edition

Editor: Jennifer Lombardo
Designer: Rachel Rising

Activity on page 5 by Science Buddies (September 26, 2019); page 11 by Science Buddies/Svenja Lohner (February 14, 2019); page 17 by Science Buddies/Ben Finio (January 24, 2019); page 23 by Science Buddies/Svenja Lohner (December 13, 2018); page 29 by Science Buddies/Ben Finio (November 8, 2018); page 35 by Science Buddies/Ben Finio (September 20, 2018); page 41 by Science Buddies/Sabine de Brabandere (June 14, 2018); page 47 by Science Buddies/Ben Finio (May 3, 2018); page 51 by Science Buddies/Ben Finio (March 8, 2018); page 57 by Science Buddies/Ben Finio (March 23, 2017).

Photo Credits: pp. 3, 4, 5, 8, 11, 15, 23, 29, 31, 35, 37, 41, 44. 45, 47, 50, 51, 53, 54, 57, 59, 60 cve iv/Shutterstock.com; pp. 4, 5, 11, 17, 23, 29, 35, 41, 47, 51, 57 Anna Frajtova/Shutterstock.com.

All illustrations by Continuum Content Solutions

Cataloging-in-Publication Data
Names: Scientific American.
Title: Gravity's pull: 10 fun gravity experiments / Scientific American.
Description: New York : Scientific American Educational, 2023. | Series: Bring science home| Includes bibliographic information, glossary and index.
Identifiers: ISBN 9781684169665 (pbk) | ISBN 9781684169672 (library bound) | ISBN 9781684169689 (ebook)
Subjects: LCSH: Experiments-- Juvenile literature | Gravity-- Juvenile literature | Science projects-- Juvenile literature
Classification: LCC QC176.3 S35 2023 | DDC 531.14 --dc23

Manufactured in the United States of America

CONTENTS

⚛ THESE ACTIVITIES INCLUDE SCIENCE FAIR PROJECT IDEAS.

INTRODUCTION

As children, we quickly learn what gravity does—even before we know what it's called. However, there's more to gravity than just making things fall down! The experiments in this book will show you how to use science to either overcome gravity or make it work for you.

Projects marked with include a section called Science Fair Project ideas. These ideas can help you develop your own original science fair project. Science fair judges tend to reward creative thought and imagination, and it helps if you are really interested in your project. You will also need to follow the scientific method. See page 61 for more information about that.

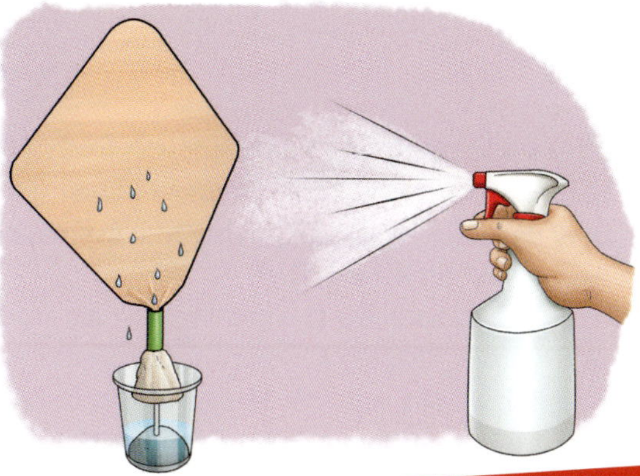

Catch Water from the Air

GET FRESH WATER—OUT OF THIN AIR! USE GRAVITY TO HARVEST "FOG" TO SEE HOW SOME PEOPLE ARE CAPTURING WATER.

How long do you think you could survive without drinking any water? Only three to four days! Your body needs water to function properly. We also rely on water for cooking, cleaning, and many other activities. You might not think about getting enough water if you live in an area where you can get it easily by turning on your faucet. However, what about areas that don't have a reliable water supply? In this activity, you will learn about one creative way to collect water—from the air!

PROJECT TIME

30-45 minutes

KEY CONCEPTS

Physics
Water
States of matter
Gravity
Weather

BACKGROUND

Many people around the world struggle to have enough fresh water for daily necessities. Because of that, people have gotten creative in finding other ways of getting the fresh water they need. For example, people who live near the coast sometimes collect water by harvesting it from fog!

Fog is a low-hanging cloud that touches the ground. The air around us contains moisture in the form of water vapor. Usually a cloud (or fog) forms when the air temperature gets cool enough that the water in the air starts to condense, which means that the water vapor turns into tiny water droplets in the air. The collection of these tiny water droplets makes a cloud (or fog) visible to us. In the case of fog, they can even make it difficult to see very far.

Coastal regions frequently get fog because the warm air of the land meets the cooler air from the ocean—right around or just above ground level. When these air layers collide, the water vapors condense and fog is created.

How can you get water out of the fog? You have to find a way to collect the tiny water droplets out of the air. This is done with big meshes that are set up perpendicular to the path of the wind. As the wind carries the fog through these meshes, the water droplets get caught by the mesh. Once the droplets accumulate and become big enough, gravity pulls them down the mesh and into containers that are set up to collect the water. In this activity, you will build your own fog catcher and harvest water from the air—with simulated fog.

MATERIALS

- Work area that can get wet
- Pantyhose
- Wire coat hanger
- Tape
- Clear drinking glass, cup, or jar
- Modeling clay
- Refillable spray bottle
- Water
- Paper
- Pen or pencil
- Scale (optional)
- Second glass, cup, or jar (optional)
- Measuring cup (optional)
- Humidifier (optional)
- Various other mesh materials (optional)

PREPARATION

- Take the wire coat hanger and form it into a diamond shape by stretching it out.
- Place one end of the pantyhose over the diamond-shaped hanger and secure it at the bottom with tape.
- Place a large lump of modeling clay at the bottom of your clear glass, cup, or jar.
- Straighten the hook of the wire coat hanger.
- Place the straightened hook into the clay so it stands upright and is stable.
- If you're using a scale, you can weigh your fog catcher and write down the weight.
- Fill the spray bottle with tap water.

PROCEDURE

- Take the spray bottle and point it away from your fog catcher. Squeeze the trigger until it starts spraying. *What can you see when the bottle starts spraying?*
- Now hold the spray bottle about 1 foot (30.5 cm) away from, and perpendicular to, the net of your fog catcher. Aim the spray bottle at the fog catcher and pump it about 10 times. Inspect the mesh afterwards. *What do you see?*
- Spray the net with water another 10 times. *What do you observe now?*
- Repeat this step about 10 times. After every 10 sprays, inspect the mesh of your fog catcher. *How does the surface of your net change every time?*
- *What do you see in the bottom of your collection container?* If you used a scale, you can place the whole fog catcher onto the scale again and write down its mass. *Did the mass of the device change? If yes, how? Can you explain the change?*

SCIENCE FAIR IDEA

Spray your sprayer directly into another glass, cup, or jar the same number of times you sprayed it onto your fog catcher. Compare the amount of water in each container or measure them in a measuring cup. *How efficient was your fog catcher at capturing the water available?*

SCIENCE FAIR IDEA

Instead of using a spray bottle, you can also place the fog catcher close to a humidifier. *How does this change your results?*

SCIENCE FAIR IDEA

Try out other mesh materials that you have in your house. *What are the best materials to catch water from the air?* You can measure the fog catching efficiency by measuring the weight of the fog catcher before and after spraying it with water. In order to compare your results, you will need to use the same amount of water for each material you test.

SCIENCE FAIR IDEA

Does it matter how you hold the spray bottle in front of the fog catcher? How does the distance or angle at which you are directing the mist toward the mesh affect the efficiency of the fog catcher?

OBSERVATIONS AND RESULTS

The device that you built from the pantyhose and coat hanger is a simple model of a fog catcher, which usually consists of a mesh stretched across a frame and a collection container below it. You used the spray bottle to mimic the fog. The nozzle inside the spray bottle breaks the stream of water into many tiny water droplets. You can see these water droplets coming out of the spray bottle. These droplets mimic fog, which is also small water droplets dispersed in the air.

When these water droplets passed through the fog catcher, they were collected by the mesh. You probably didn't see much after 10 sprays, but after awhile, you should have seen tiny little water droplets building up on the surface of the pantyhose. The more fog the mesh collected, the more water droplets accumulated. At some point, they probably started to merge and form bigger drops that then ran down the surface into the container—thanks to gravity. You could probably start to see the water pooling in the bottom of your container. If you used a scale, you could weigh the difference between the device at the beginning of the activity and at the end, which reflects the amount of water you collected from the air!

If you tried different meshes, you might have collected different amounts of water. Scientists are still trying to optimize the materials and designs of fog catchers so they can be even more efficient in water harvesting. Real-life fog catchers, depending on the mesh material and design, can harvest as much as 2.4 gallons (9 liters) of water per day for each square meter of mesh.

CLEANUP

If the pantyhose are still intact after the activity, you can dry and reuse them. You can also reshape the wire hanger into hanger form and reuse it as well.

Slippery Slopes, and the Angle of Repose

PILE IT ON! LEARN HOW PHYSICS EXPLAINS WHY DIFFERENT MATERIALS CREATE DIFFERENT SHAPED PILES.

PROJECT TIME

60-120 minutes

KEY CONCEPTS

Physics
Gravity
Forces
Angle of repose

Have you ever seen video footage of an avalanche or landslide rolling down a hill? Why is it that at one moment everything seems fine, then suddenly the mountain begins to slump? This movement has something to do with how the snow or soil is piled up on the mountain. Granular materials such as snow or soil generally pile up relatively well. If the slope angle gets too steep, however, the materials will start to slide down the slope. This critical slope angle, also called the angle of repose, is different for different materials. In this activity, you will create your own small avalanches and determine the angle of repose for different materials along the way!

BACKGROUND

If you pour a granular material on a flat surface, it will form a conical pile. If you add more of the material, the pile will grow. At some point, however, the slope angle of the pile will always stay the same. This is because as the pile grows and its slope reaches a certain angle, some material will slide down the pile. This is the angle of repose and is the steepest angle at which a material can be heaped without sliding down. However, why would the material slide?

The reason is gravity. The gravitational force acting on the material on the slope can be split into two different components. One, the normal force, pulls the material into the slope in a direction perpendicular to the slope surface. The normal force pulls inward on the grains on the slope, which actually helps hold the grains together and prevents the material from sliding downward. Depending on the type and shape of material, frictional forces between the grains might also hold them together. As a result, grains of irregular shape that have the ability to interlock tend to have a higher angle of repose. The second gravitational component is the shear force, which pulls the grains down the slope in a direction parallel to the slope's surface. The steeper the slope, the higher the shear force will be. At some point, the shear force will overcome the normal force of gravity. This is usually the moment the materials start sliding down the slope and the angle of repose is reached.

This might sound like a very theoretical concept. There are plenty of situations, however, in which grains such as corn, flour, or gravel need to be piled up. In these situations, knowledge of their angle of repose can be very helpful in figuring out the proper dimensions of a storage silo or to design the right-size conveyor belt to transport them. The angle of repose is also used to assess whether a mountain slope is going to collapse. This helps geologists or mountaineers know the risks of avalanches ahead of time! There are several ways to measure the angle of repose of a specific material. One, which you will be doing in this activity, involves measuring the height and radius of a pile formed by a material, then using these numbers to calculate the angle of repose.

MATERIALS

- Adult helper
- Scissors
- Disposable plastic cup, 16 ounces
- Printer paper
- Baking dish
- Pen
- Ruler

- Tape measure
- Table salt (at least one cup)
- Rice (at least one cup)
- Powdered sugar (at least one cup)
- Scientific calculator or smartphone app with scientific calculator functions
- Additional granular materials such as lentils, flour, etc.
- Protractor (optional)

PREPARATION

- With the help of an adult, cut a small hole in the bottom of the plastic cup. Its diameter should be about 0.8 inch (2 cm).

PROCEDURE

- Place a sheet of printer paper into the baking dish and label it with the material that you want to test.
- Covering the cup's hole with your hand or fingers, fill the cup at least halfway with your first material.
- Hold the cup close to the top of the printer paper at its center. Then remove your hand or fingers to release the material inside the cup. *What do you notice when the material falls on the paper?*
- As the material pile grows, hold your cup higher so all the material can fall onto the paper. Depending on the material, you might need to tap the cup a little to help it all come out. Observe the slope angle of your pile as it grows. *Does the slope angle change over time? How do the material pile and slope angle look at the end?*

- With a pen, carefully draw along the circumference of the material pile. Be careful not to disturb the pile too much. *How big or small is your circle?*

- With the ruler, measure the height (h) of the material pile in centimeters. You can very carefully slide the ruler into the pile to measure its height at its peak. If this method disturbs the pile too much, you can also hold the ruler next to the pile and carefully extend a tape measure from the top of the pile to the ruler. The height of the pile is where the tape measure and ruler intersect. Write down the height of the pile on the sheet of paper next to the pile. *How tall did your material pile get?*

- Using your drawn circle and the measured height, calculate the angle of repose for this material. The equation for calculating the angle of repose is tan-1(h/r). Don't worry if the equation looks complicated—you will determine each number step by step, and the rest is done by the calculator!

- Remove the material from the sheet of paper. Using the ruler, measure two different diameters (d) of the drawn circle in centimeters. To do this, draw two lines from one random edge of the circle through its center to its opposite edge. The length of each line will give you the circle's diameter. Write down both numbers. *Are the numbers very different? What does this tell you about the shape of the circle?*

- Calculate the average diameter of your circle by adding both measured diameters and dividing the result by two. From this you can calculate the radius (r) of your circle by dividing the average diameter by two again.

- Use the calculator to divide the measured height (in centimeters) by the calculated radius (in centimeters). Write down the result to one decimal point.

- Now the only step left is to enter this number into the calculator and hit the inverse tangent key (or tan-1). This will give you the angle of repose. Write it down on your sheet of paper.

- Repeat these steps with all the other materials. *How does the shape and size of each material pile differ? Which material has the lowest or highest angle of repose? Did you expect these results?*

SCIENCE FAIR IDEA

Test some more materials. Any granular material will work. Other possibilities to test are sand, coffee beans, gravel, cornstarch, etc. *How do these materials compare with the ones you have tested?*

SCIENCE FAIR IDEA

Use a protractor to measure the angle of repose directly from the pile you created with each material. *How well does your measured angle match up with your calculated angle of repose?*

OBSERVATIONS AND RESULTS

How did your piles look? Each of your materials should have formed a nice conical pile. The circumference of each pile should have been close to a symmetric circle, which means the two measured diameters should be relatively similar. The heights of the piles and the sizes of the measured circles, however, should have changed depending on the materials you tested. The rice probably formed large circles, whereas powdered sugar likely resulted in a very small circle. Conversely, the height of the rice pile should have been significantly lower than that of the powdered sugar.

Based on these numbers, you probably found rice has a small angle of repose (around 25 to 30 degrees), whereas the powdered sugar has a relatively high angle (greater than 40 degrees). The calculated angle of salt should be somewhere in between those. This variation is due to the different sizes and shapes of the material particles. Increasing particle size will generally decrease the angle of repose. This is why the large rice particles have a much lower angle than that of the fine-grained powdered sugar. Additionally, particles that are irregularly shaped hold together much better than particles that are very round and easily roll over each other.

CLEANUP

If you used clean materials, you can reuse the rice, salt, and powdered sugar. Clean your work area and wash your hands.

Perform a Scientific Balancing Act

DEFY GRAVITY WITH A LITTLE PHYSICS!

What makes an object stay balanced? Look around you. Most of the objects in the room are probably balanced—and not on the verge of tipping over. If someone hands you an object and asks you to put it down, you probably intuitively know how to place it so it won't fall over. However, what's the science behind how an object balances? Why do certain objects only balance on some sides and not others? Try this project to find out!

PROJECT TIME

45-60 minutes

KEY CONCEPTS

Physics
Center of mass
Gravity
Area

BACKGROUND

In general, we use the word "balanced" to refer to an object that is upright and not falling over. The technical term for an object that won't tip over—even if it is pushed—is stable. An object that can be knocked over by a light push or a gentle puff of wind is unstable. A chair sitting on the floor on all four legs, for example, is stable—it's hard to knock it over. If you try to balance the chair on one leg, however, it's unstable. The moment you let go of the chair, it will probably fall.

What determines if an object is stable or unstable? It depends on two things: the location of the object's center of mass and where the object is in contact with the ground. For a perfectly symmetrical object (such as a ball), the center of mass is directly in the middle. For an irregularly shaped object (such as a chair), the location of the center depends on how the object's mass is distributed. Imagine looking down on a chair from directly above and drawing imaginary lines connecting its four legs, forming a square on the ground. The chair's center of mass will be inside this square. Gravity pulls down on the chair, acting like a single force concentrated at the center of mass. Because this force acts inside the chair's contact area with the ground, it does not cause the chair to tip over.

When you try to balance the chair on one leg, conversely, it has a very small contact area with the ground. It's almost impossible to get the chair's center of mass to line up inside this contact area—and even if you do, the slightest motion will move it back outside. This will cause the chair to tip over because the force exerts a torque on the chair.

In this activity, you'll try this simple test with a variety of objects of different shapes. Do you think you can balance them all?

MATERIALS

- An assortment of objects from around the house (that you have permission to use) to try to balance. Get a variety of different-shaped objects, such as: rectangular objects (blocks, boxes, etc.); long, skinny objects (pencils, rulers, etc.); and irregularly shaped objects (coffee mugs, hairbrushes, etc.).
- Helper (optional)

PREPARATION

- Gather your objects on a flat, firm surface on which they can be balanced.

PROCEDURE

- Take one of your rectangular objects. Predict which sides you think it will balance on, then try balancing it on each side.
- Now try balancing it on the edges and corners. *Can you still get it to balance?*
- Now take one of your long, skinny objects. Predict which sides you think it will balance on. First try to balance it on its longest side.
- Now try balancing it on one of the skinny sides or points. *Can you still get it to balance?*
- Try one of your irregularly shaped objects. Again, first predict how you think it will balance, then try it out. You can probably get a coffee mug, for example, to balance easily on its bottom, flat side. *Can you get it to balance on its rounded side or handle?*

EXTRA

After you get an object to balance, test whether it is stable or unstable. Gently blow on it or nudge it with your finger. *Does it fall over or remain standing?*

EXTRA

Get a helper and try this test with larger objects, such as a chair or broomstick. Be careful not to drop anything heavy!

EXTRA

Try balancing objects on your fingertip instead of on a flat surface. *How do your results change?*

OBSERVATIONS AND RESULTS

You probably found it was easier to balance objects on larger, flatter sides than on smaller, pointier, or rounder sides. It's also easier to balance an object that has three or more points of contact with the ground (like the legs of a chair). This is because when an object has a large, flat surface in contact with the ground (or multiple points forming a polygon), most of the time its center of mass lies inside this area and the object is stable. Thin edges, points, and round surfaces form very small contact areas with the ground. It's difficult to get a center of mass to stay inside this size area, making the object unstable. You might be able to temporarily balance an unstable object—for example, balancing a pencil vertically on its eraser—but it only takes a gentle push to knock the object over.

CLEANUP

Put away the materials you used.

21

A Candle Seesaw Balancing Act

STRIKE UP SOME PHYSICS FLASH WITH THIS IMPRESSIVE ACTIVITY.

When the days grow shorter and it gets dark early in the evening, many people enjoy candlelight. Candles are also a great tool for doing science—so why not combine both and add a little light to your science? In this activity, you will use candles to investigate the balancing forces of a seesaw.

PROJECT TIME

60-75 minutes

KEY CONCEPTS

Physics
Force
Gravity
Mass
Lever

BACKGROUND

Seesaws are classic playground equipment for children. They are not only fun to play on but also provide an excellent opportunity to explore a type of simple machine designed to lift objects much heavier than your strength would normally allow. A seesaw is a specific type of lever; it consists of a long beam attached to a pivot called the fulcrum. As soon as you put weight on one end by sitting on one side of the beam, it drops to the ground. This is because the force of gravity is acting on the weight of your body, pulling it and the beam down. How much weight is pulling down on the beam depends on your body mass. The heavier you are, the larger the gravitational force. To balance the beam again, you need a counteracting force on the other side. One possibility is to place a second person of the same weight on the other side of the beam. Once the same force is pulling down on each side of the beam, the seesaw is balanced.

You might know from experience, however, that it is not only weight that matters but also where people are seated along the beam. Two people with the same body mass sitting on opposite sides of the beam can still change the balance and make the seesaw rotate by moving farther from or closer to the beam's center. The beam's rotation is caused by the turning force, also called torque, which takes into account the force pulling down on the beam and location of the applied force.

According to the law of the lever, the seesaw will stop rotating once the turning forces are equal on both sides and cancel each other out. This law also explains why you are able to lift very heavy objects with a lever. When the object on one side is too heavy for you to lift, you just have to move farther away from the beam's center until your turning force is greater than the one on the other side.

Who would have thought such a simple activity would entail such complex physics? Don't worry if this sounds complicated; it will become more obvious once you try this project—using a seesaw made of candles!

MATERIALS

- Two identical birthday candles
- Strong tape
- Needle that is longer than the candle's diameter
- Aluminum foil
- Knife
- Two glasses of the same height
- Lighter
- Adult helper
- Safe location and surface for using candles

PREPARATION

- Tape the birthday candles together at their ends so both wicks are facing opposite directions.
- Put a large piece of aluminum foil on your work area to protect it from any wax spills.
- Set the two glasses next to each other in the middle of the aluminum foil. The gap between the glasses should be small enough to place the needle across it.
- Have an adult push the needle all the way through the side of the candle exactly where the ends of both candles meet. This should be exactly in the middle between both wicks. If it is too difficult to push the needle through the candle, try to heat the needle in a flame before you push it through the wax.

PROCEDURE

- Place the candle in the gap between the glasses so the parts of the needle that are sticking out on each side of the candle rest on the rim of each glass. *What do you notice once the candle is placed between the glasses? Can you see the similarities between your experimental setup and a seesaw? Where is the fulcrum of your candle seesaw?*
- If your candle seesaw is unbalanced, change the location of the needle in the candle. Once the needle is placed exactly in the middle of the two candles, the seesaw should be balanced. *Why does the needle have to be exactly in the middle of the candle to balance the seesaw?*
- Make sure your surface is covered with aluminum foil along the entire length of the candle.
- Once the candle is balanced and doesn't drop down on either side, ask your adult helper to carefully light both candles. Don't light both candles at the same time—wait for a couple of seconds before you light the second one. *Do you think it would make a difference if both candles were lit simultaneously?*

- Watch how both candles burn and observe the movement of your candle seesaw. *What happens with the seesaw after a while? Does it stay balanced or does it start moving? If you see movement, can you explain why the candle seesaw is moving?*
- Once the candles burn down by about one-fourth, blow out both candles and cut the top (approximately 0.5 inch, or 1 cm) off one of the candles. *How do you think this will affect the seesaw balance?*
- Place the candle between the glasses again so the parts of the needle that are sticking out on each side of the candle rest on the rim of each glass. *Is the seesaw still balanced as before? Why or why not?*
- Ask your adult helper to light both candles the same way as before.
- Watch the candles burn and observe what happens. *Does the seesaw move again? How are your results different from the previous observations?*
- Make sure to blow out both candles before they burn completely.

EXTRA

Repeat the activity, and this time don't place the needle in the middle of the candle beam. One side should be heavier, similar to when you cut the top off one candle. This time, however, only light the candle on the longer side. Then watch the candle burn. *What happens to the candle seesaw this time? Does it start moving? Can you explain your observations?*

OBSERVATIONS AND RESULTS

Did you notice that what you were building resembled a seesaw? Both candles taped together formed a long beam that was attached to the needle, which acted as the pivot (or fulcrum). The candle beam was able to rotate freely from one side to the other just like a real seesaw. Because you don't put any extra weight on the candle beam as you would on a playground seesaw, the only force pulling down on the beam is the weight of the candle itself. To balance the seesaw, it is important that both forces pulling down on each side of the beam are exactly equal. This is only true if the needle is placed exactly in the middle of the candle beam. If one side is slightly longer, this would also make it heavier and it would drop down, as you might have observed. If the needle is placed in the middle, however, the gravitational forces pulling down on each side should cancel each other out, and it should stay balanced.

This changes once you light the candles. When the candle is burning, a chemical reaction occurs that converts the candle wax to a gas. You probably also noticed that the solid wax turned into a liquid and dripped onto the aluminum foil. The wax lost through burning and dripping makes the candle shorter and therefore lighter. As this side of the candle beam becomes lighter, it moves upward, whereas the other, heavier side drops down. The rotation is reversed once the other candle loses wax and becomes lighter again. The key to this seesaw movement is that both candles are not burning the same amount of wax at the same time. This is the reason why you have to light them one after the other. If both candles start burning at exactly the same time and lose the same amount of wax simultaneously, the candle seesaw would stay balanced.

If you cut part of the candle on one side, the longer, now heavier side of the beam would drop down. When you lit the candles, you probably noticed the candle seesaw didn't move at all because the longer side will always be heavier while both candles are burning. If you did the extra test, however, and only lit up the candle on the heavier side, you would have noticed that as soon as the candle burned down enough to make it shorter and lighter, the seesaw dropped down on the other side.

CLEANUP

Make sure both candles are fully extinguished and dispose of them and the aluminum foil. If you do not want to reuse the needle, make sure to dispose of it in a sharps container.

Hula-Hooping with a Rubber Band

DO THE LOOP-DE-LOOP WITH A RUBBER BAND HULA HOOP—AND USE PHYSICS TO SEEM TO DEFY GRAVITY!

PROJECT TIME

20–30 minutes

KEY CONCEPTS

Physics
Gravity
Friction
Centripetal force
Normal force
Rotational motion
Newton's laws of motion

Are you any good at hula-hooping? If not, don't worry—you can do this fun project without any hula-hooping experience. You will examine some of the fascinating physics behind hula-hooping using just a pencil and a rubber band.

BACKGROUND

Hula-hooping is all about forces! You might not think about physics much when you play with a Hula-Hoop, but there are many different forces at work that help keep a Hula-Hoop spinning and prevent it from falling to the ground. A force is a push or a pull that acts on an object. Forces can make things move, but just because something isn't moving doesn't mean it doesn't have any forces acting on it. For example, if you're sitting in a chair right now, the chair is exerting an upward force on you that prevents you from falling to the ground. Conversely, just because something is moving doesn't necessarily mean it has force acting on it. An object moving at a constant velocity will keep moving in a straight line forever if there are no forces to slow it down (Newton's first law of motion).

So what forces act on a Hula-Hoop?

—Weight, or the force of gravity pulling the Hula-Hoop down;

—Friction, or the force that opposes motion between two surfaces that are sliding against each other (so in this case, the friction between the Hula-Hoop and your clothes);

—The normal force. In physics, "normal" means "perpendicular to," not "regular" as it does in everyday speech. The normal force is that which acts perpendicular (at a right angle) to two surfaces that touch each other. For example, a book sitting on a table has a normal force from the table pushing it up, which prevents it from falling. If you push your hand against a wall, the wall exerts a horizontal normal force against your hand (Newton's third law of motion);

—The centripetal force, which is the force that keeps a rotating object moving in a circle instead of flying off in a straight line. Imagine, for example, twirling a rock on a string. The tension in the string pulls on the rock and makes it move in a circle. If you suddenly cut the string, there is no more centripetal force acting on the rock, and it will fly away instead of continuing to move in a circle. In a Hula-Hoop, the centripetal force results from your body pushing on the hoop (a combination of the frictional and normal forces).

As you will see, the combined effect of all these forces determines the Hula-Hoop's motion (Newton's second law of motion).

MATERIALS

- Pencil
- Rubber band

PREPARATION

- Hold the pencil by the eraser end and point the tip up.
- Loop the rubber band over the pencil so it falls down to your fingers.

PROCEDURE

- Slowly start twirling the pencil. *What happens?*
- Keep twirling the pencil faster and faster. *How fast do you have to twirl it before the rubber band starts moving up?*
- Stop twirling. *What happens?*
- Try to twirl the pencil so fast that you get the rubber band to fly off the tip. *Can you get it to work?*
- Try changing the angle of the pencil as your twirl it. This means you do not keep the pencil perfectly vertical as you twirl it. You can pinch the eraser end with your fingers and spin the tip around in a circle, tracing out a three-dimensional cone. *What happens if you make this "cone" wider? Does it make it easier or harder to make the rubber band move upward?*
- Try rotating your wrist as you twirl the pencil so the pencil becomes horizontal and eventually upside down (so the tip points downward). *Can you keep the rubber band on the pencil even when it's upside down?*

EXTRA

Try using different size rubber bands. *What's easier to twirl, a big or a small one?*

OBSERVATIONS AND RESULTS

In this activity, the pencil acts like a person's body and the rubber band acts like a Hula-Hoop. Just like with a real Hula-Hoop, you should have found that if you did not spin the pencil fast enough, the rubber band would fall down. As you spin the pencil faster and/or make the cone you trace with the pencil wider, the rubber band should start to move up the pencil and eventually fly off the tip!

To understand this, you can refer back to the information in the Background section. Remember the example about centripetal force and twirling a stone on a string? The stone "wants" to keep moving in a straight line, but the centripetal force from the string makes it move in a circle. The rubber band behaves in a similar manner. It "wants" to fly off in a straight line, but the centripetal force from the pencil makes it move in a circle. If you spin the pencil fast enough, the centripetal force is no longer strong enough to hold the rubber band in place—so it starts sliding outward (and upward). This makes it very difficult, if not impossible, to keep the rubber band on the pencil when you hold it upside down.

CLEANUP

Put the pencil and rubber band back where you found them.

Lights, Camera, (Capillary) Action!

SLURP! LEARN HOW TREES CAN SIP WATER FROM THE GROUND—ALL THE WAY UP TO THEIR LEAVES—WITH A LITTLE AT-HOME PHYSICS.

PROJECT TIME

30-45 minutes

KEY CONCEPTS

Physics
Adhesion
Cohesion
Surface tension
Gravity

How do trees suck water all the way up to their leaves? How do paper towels soak up a spill? Are these things related? Try this project to learn about capillary action, and repeat a classic demonstration from over 100 years ago!

BACKGROUND

Have you ever looked closely at water in a drinking glass? You might notice the surface of the water is not completely flat; rather, it forms a small lip, called a meniscus, that curls up around the edge of the glass. This occurs because of two forces: adhesion (the attraction between the water molecules and the glass) and cohesion (the attraction of water molecules with one another). Cohesion among water molecules gives rise to surface tension, or the way water's surface acts like a stretchy "skin." Surface tension helps prevent raindrops from breaking apart and allows insects such as water striders to walk on water without sinking. It also contributes to the water rising along the edge of the glass: As some water molecules are pulled up because of their attraction to the glass, they pull others on the surface along with them.

This phenomenon, called capillary action, allows water to be sucked up into small gaps, seemingly defying gravity. This might not seem like a big deal for a meniscus that's only a couple millimeters high in a glass of water. However, what about a paper towel sucking water up a few inches or a tree pulling it up tens or even hundreds of feet? How can they possibly get the water up so high? In this activity, you will investigate how the size of the gap affects the height to which the water will rise.

MATERIALS

- Two small picture frames with glass covers
- Two paper clips
- Rubber band
- Tray or plate
- Water
- Food coloring
- Dish towel

PREPARATION

- Gather your materials on a work surface that can tolerate spills of colored water.
- Fill a tray or plate with a shallow layer of water and mix in a few drops of food coloring.
- Remove the glass plates from the picture frames.

PROCEDURE

- Use the rubber band to hold the two glass plates together, with the two paper clips in between opposite edges as spacers. *Do you think you can get water to defy gravity?*
- Dip one edge of the glass plates (a side without a paper clip) into the water. Hold the plates still and wait a few seconds. *What happens?*
- Separate the glass plates and wipe them dry with a towel.
- Use the rubber band to hold the plates together again, but this time, only use one paper clip on one side (so at the opposite edge, the glass plates should be touching each other).
- Dip one edge of the glass plates (one of the sides adjacent to the side with the paper clip, not opposite it) into the water again. Hold the plates still and wait for a few seconds. *What happens this time?*

SCIENCE FAIR IDEA

Try modifying the parameters of the tests. For example, use larger or smaller plates of glass; a thicker or thinner object as a spacer instead of a paper clip; a different liquid; etc. *How do the results change?*

OBSERVATIONS AND RESULTS

When you put two paper clips between the glass plates as spacers, the width of the gap between them was the same everywhere. When you dipped the edge of the glass plates into the water, it was sucked up into the gap by capillary action (and the food coloring made this easier to see). Because the gap was the same width everywhere, the water should have risen to about the same height along the length of the plates.

When you remove one of the paper clips, something interesting happens: The gap is now wider on one end and gradually gets narrower until it disappears and the glass plates touch. This time when you dip the plates in the water, it goes up higher as the gap gets narrower. The resulting shape is called a hyperbola. This demonstrates how capillary action can lift water up higher in narrower spaces. As the gap gets wider, the weight of the water increases faster than the force available to pull it up, thus the water cannot rise as high. So if you want to suck water up very high (such as all the way into a tree), you need a very narrow gap!

CLEANUP

Use a towel to clean any spilled water and wipe the glass plates dry.

Parachutes with Holes

LOOK OUT BELOW—EVENTUALLY! TEST OUT WHAT FACTORS MAKE AN EFFECTIVE TOY PARACHUTE, AND LEARN SOME PHYSICS ALONG THE WAY.

Have you ever wondered what a parachute and an open rain jacket have in common? They both trap air and slow you down when you move fast! In this activity, you will design a parachute for a miniature action figure. Tissue paper or a plastic bag and a few strings are all it takes to make your figure into an expert skydiver.

PROJECT TIME

60–75 minutes

KEY CONCEPTS

Physics
Gravity
Drag
Air resistance

BACKGROUND

Things fall because gravity pulls them down—and the faster they fall, the harder they land. A lot of people think that heavier objects fall faster. Galileo—a Renaissance philosopher and scientist—showed that this idea, although intuitive, is wrong. You can test this idea, too. Drop a small square of cardboard (for example, 3 inches by 3 inches [8 cm by 8 cm]) and piece of paper folded to the same size simultaneously from the same height. Observe that although the paper is lighter than the cardboard, they both reach the ground at the same time. Repeating this with the unfolded sheet of paper will illustrate why parachutes work. The paper falls more slowly when it is in a larger sheet. Why? The explanation is in the air.

The air around us is made of small particles. Just like you move water particles out of the way when you pass through water, you push air particles out of the way when you move through air. Maybe you have felt how you pushed air out of the way on a bike ride. As you push the air, it pushes at you. It slides by you and feels like wind. This is called air resistance, or drag—and it slows you down. You might not like air resistance when you are on your bike, but it is ideal when it comes to slowing down a fall! The unfolded sheet of paper fell more slowly than the folded piece of paper because its larger flat surface needed to push more air away. It experienced more drag. Are you wondering how this concept can help you create the best parachute? Try the activity to find out!

MATERIALS

- Tissue paper or a plastic bag
- Scissors
- Ruler
- Tape
- Hole puncher
- Twine
- Small, nonbreakable action figure or miniature doll that may be dropped on the floor (If you do not have an action figure, use a piece of clay, a small wooden block, a measuring spoon, etc.)
- Safe location to drop your parachute from—for example, a 2nd-floor window, balcony, or open staircase

PREPARATION

- To make the canopy for the parachute, cut a square out of the tissue paper (or plastic bag). Make each side of the square 1 foot (31 cm) long. Reinforce the corners with tape and punch a hole in each corner.
- Create suspension lines by cutting four strings from the twine, each 1 foot (31 cm) long.
- To assemble your parachute, attach one end of each suspension line to each corner of the tissue paper, fold the canopy in four so its corners lay on top of one another, and knot the unattached ends of the four suspension lines together.

PROCEDURE

- Bring your parachute and action figure to your test location.
- In a moment, you will drop your figure (without the parachute) from this spot. *Do you think it will make a soft or hard landing? Can you predict with precision where it will land?*
- Drop your figure. *Were your predictions correct?*
- Drop it from the same location several more times. *How would you describe these falls?*
- Attach the parachute to your figure by winding the knotted end of the suspension lines around the middle of your figure and securing it with a knot or tape.
- In a moment, you will drop your figure (equipped with a parachute) from this spot. *How do you think it will land this time? Can you also predict where it will land?*
- Fold your canopy in four so its four corners lay on top of one another. Make sure the suspension lines are not tangled. Pick the parachute up from the corner diagonally opposite the corner with the strings. Your figure should now hang under the parachute.
- Drop your figure equipped with the parachute. *Were your predictions correct?*

- Drop it several more times. *How are these falls different from the drops without a parachute? Why would a parachute create these differences?*

- *What do you think will happen if you make a hole in the middle of the parachute?*

- Fold your parachute in four so the corners are stacked. Cut the tip of the corner that is diagonally opposite the corner with the strings attached. Open your parachute and see that there is now a hole in the middle of the canopy. *How do you predict this will affect the fall?*

- Fold your canopy in four again. Pick it up at the corner that has been cut away and drop your figure. *Was your prediction correct?*

- Drop it several more times. *How did the hole change the way the figure falls? Why would this happen?*

SCIENCE FAIR IDEA

Investigate how many holes you can create before your parachute no longer functions. Experiment with placement to see whether or not the location of the hole makes a difference. You could also gradually increase the size of the hole and study how its performance changes.

SCIENCE FAIR IDEA

Make canopies of different materials, sizes, and shapes. *Which ones work best?*

SCIENCE FAIR IDEA

Measure the impact of the fall by letting your figure land in a sandbox. How deep is the indent created by the fall?

SCIENCE FAIR IDEA

Use a timer to measure how long the fall takes. *Can you calculate the average speed of your figure during the fall? Which parachute creates the slowest fall?*

OBSERVATIONS AND RESULTS

Your figure probably fell straight down and had a hard landing at first. The figure equipped with a parachute probably had a softer landing, but it was probably also harder to predict where exactly it would land. The figure equipped with the parachute with the hole in the middle probably still had a pretty soft landing.

Gravity pulls objects straight down toward the center of Earth. It is strong enough to make falling objects move quickly, creating hard landings! Luckily we have a layer of air around our planet that slows falls. Air resistance, or drag, pushes against objects when they fall. Parachutes catch a lot of air, creating a lot of drag. They can drastically slow a fall, allowing a softer landing. This slow drop, however, can be hard to control. A figure landing with a parachute might sway to the side during the fall.

Some parachutes trap air, just like a loose jacket can trap air on a bike ride. This trapped air wants to escape. It can often only escape at the edges, which makes those edges (canopy edges or the sides of your jacket) flap. Some parachutes have a hole in the center to release air in a controlled way. It makes the parachute more stable, with only a minimal change in drag.

CLEANUP

Put away any materials that can be reused. Throw out or recycle any scraps.

Walking Water

CAN YOU FILL AN EMPTY CUP—WITHOUT POURING ANYTHING INTO IT? FIND OUT HOW PHYSICS CAN HELP YOU DEFY GRAVITY!

PROJECT TIME

12-24 hours

KEY CONCEPTS

Physics
Gravity
Molecules
Surface tension

Imagine this challenge: You have two glasses of water—one empty and one full. You want to pour half of the full glass into the empty one. The twist? You aren't allowed to pick up either glass! Can you get the water to "walk" between the glasses using nothing but a paper towel? Try this activity to find out!

BACKGROUND

You've probably used paper towels to clean up spilled water or other liquids. Have you ever wondered exactly how they soak up so much water? Paper towels are made of many small fibers that have gaps in between them. Water gets pulled into these gaps by capillary action—the same phenomenon that allows trees to suck water out of the ground. This action is partially fueled by surface tension, which is caused by cohesion (water molecules being attracted to one another). Surface tension is what allows water to form beads instead of spreading out, and for some small insects to walk on water. It also allows water to get sucked up into narrow tubes or gaps in materials. The absorption process is also aided by adhesion (the attraction between different types of molecules). The paper towel fibers are made of cellulose, which also comprises wood and many plants. These fibers are polar, meaning they have a slight positive charge on one end and a slight negative charge on the other. Water molecules are also polar. Because opposite electric charges are attracted to each other, this results in the water molecules being attracted to the cellulose fibers.

Although you might not have tried it yourself, maybe you've heard of another phenomenon: siphoning. This process uses a tube to suck liquid from one container into another, somehow seeming to defy gravity. The liquid gets sucked up into the tube before it moves down into the next container—even though there's no pump powering the motion. Siphoning is based on some of the same physical principles mentioned above. (For example, water can be drawn up into the tube by capillary action.)

That might sound like a lot of science for a simple, everyday object. Try this activity to find out if you can use paper towels to move water from one cup to another—and understand the science behind it!

MATERIALS

- Odd number of clear glasses or cups (at least three)
- Water
- Food coloring
- Spoon
- Half-sheet paper towels (at least three)

PREPARATION

- Line up all your glasses in a row.
- Starting with a glass on one end, fill every other glass with water.
- Put a few drops of food coloring in each water-filled glass. You can choose what colors to use, but don't use the same color twice in a row.
- Use the spoon to mix the food coloring in each glass. Use a paper towel to wipe off the spoon in between glasses so you don't transfer the colors.

PROCEDURE

- Fold each half-sheet paper towel (except the one you used to clean the spoon) into a narrow strip about 1 inch (2.5 cm) wide.
- Fold each paper towel in half lengthwise to form a "V" shape. The V should be only slightly taller than your glasses. If necessary, rip or cut a little bit off each end of the V to make it shorter.
- Use one paper towel to connect each pair of adjacent glasses. (Flip the V shape upside-down and put one end in each cup.)
- Look closely at the ends of the paper towels that are in the glasses with water. *What do you notice?*
- Take a break! This experiment goes very slowly. Come back in 15 or 20 minutes. *What do you see now?*
- Keep checking on your setup over the next couple of hours. *What happens?*
- Let your test sit overnight, and check on it the next day. *What does it look like now?*

SCIENCE FAIR IDEA

Try the activity with different brands of paper towels. *Do some work better or faster than others?*

SCIENCE FAIR IDEA

Try placing your cups in different arrangements instead of a straight line. *For example, what happens if you connect three cups, initially filled with water, to an empty fourth, central cup?*

SCIENCE FAIR IDEA

What happens if you use a paper towel to connect two cups that are initially filled to different heights with water?

OBSERVATIONS AND RESULTS

As soon as you place the paper towels in the glasses, you should notice they start to absorb some of the water. The water starts getting sucked up into the paper towel due to capillary action (described in the Background section) and eventually starts going down the other side into the empty glass (just like a siphon). This process happens very slowly though—it's like watching paint dry or grass grow! You might need to go do something else for a couple hours before you can start seeing water accumulate in the empty glasses. The water will eventually stop flowing when the water level in all the cups is even—but you will probably have to let it sit overnight before this happens.

When two different colors of water mix, the food-coloring dyes combine to form a third color.

CLEANUP

Wash and put away the glasses. Throw out the wet paper towels.

The Physics of Bottle-Flipping

DON'T FLIP OUT—IT'S PHYSICS!

The bottle-flipping craze might be dying down, but it isn't too late to investigate the physics of this internet sensation. Even if you've never heard of it, give this project a try—not only can you impress your friends with a fun new trick, but you'll also be able to explain the science behind it!

PROJECT TIME

30–45 minutes

KEY CONCEPTS

Physics
Mass
Gravity
Angular momentum

BACKGROUND

"Bottle flipping" took the internet by storm in 2016. If you haven't seen it yet, search for it on YouTube or your favorite social network; you're bound to find a few videos. The process involves flipping a partially filled water bottle into the air so it lands upright. This might seem like a very simple concept, but the physics behind it is actually quite complex—and it takes some practice to master the feat!

To understand the physics of bottle-flipping, first you need to understand angular momentum. An object's angular momentum depends on its angular velocity (how fast it is spinning) and its moment of inertia (how much its mass is spread out from a central point). When no external torque acts on an object, its angular momentum must be conserved. The classic example of this is a spinning ice skater. If she is first spinning with her arms extended, she has a high moment of inertia (her mass is spread out, away from the center of her body). If she pulls her arms in tightly, her moment of inertia decreases. In order for her angular momentum to stay the same, her angular velocity must increase so she spins faster. You can observe this for yourself in a spinning office chair.

What does that have to do with bottle-flipping? Imagine throwing a rigid object, such as a coin. Gravity will pull the coin back down to the ground. Because the object is solid, the distribution of its mass does not change as it flies and spins through the air, and its moment of inertia and angular velocity remain the same. That makes it very difficult to predict whether the coin will land with heads or tails up because it keeps spinning as it falls. A water bottle is different, however. It contains liquid water, which is free to slosh around inside the bottle, changing the distribution of mass. Just like an ice skater spreading out or pulling in her arms, this changes the bottle's moment of inertia and therefore its angular velocity (because the total angular momentum must stay the same). You can exploit this fact to make it easier to successfully flip a bottle. How? Try this activity to find out!

MATERIALS

- Plastic water bottle
- Tap water

PREPARATION

- If you have never tried bottle-flipping before, you should practice before you start this project. You want your technique to remain consistent (for example, how high you throw the bottle, how far you throw it horizontally, and how fast you spin it) throughout the activity.

- Fill a plastic water bottle about one-quarter to one-third full with water and put the cap on tightly.
- Hold the bottle loosely by the neck and toss it forward (so the bottom rotates away from you).
- Try to throw the bottle so that it does one complete flip and lands upright without falling over. This can take a lot of practice!
- If you get frustrated, at least try to observe which side the bottle initially lands on (top, bottom, or side), even if it falls over after that. *Can you get the bottle to consistently land on its bottom?*

PROCEDURE

- Once you have practiced your bottle-flipping method, try it 10 times in a row. Remember to keep your technique as consistent as possible. *How many times can you get the bottle to land upright?*
- Now try it 10 times with an **empty** bottle. *Can you still get the bottle to land upright?*
- Now try it 10 times with a **completely full** bottle. *Can you still get the bottle to land upright?*
- Try to see if you can find the optimal amount of water in the bottle. *What if the bottle is one half or three quarters full? What amount of water gives you the best success rate?*

SCIENCE FAIR IDEA

Put some water bottles filled with different amounts of water in the freezer overnight (make sure they are sitting upright). Try flipping them the next day. *Is it easier or harder to successfully flip bottles with ice instead of liquid water inside them?*

SCIENCE FAIR IDEA

Try throwing the bottle different distances and heights—and vary how much you spin it. *Is it easier to get the bottle to land upright if you throw it across the room or so it lands just in front of you? What if you try to land it on a table instead of the floor? What if you try to get it to complete two flips instead of one?*

SCIENCE FAIR IDEA

Try landing the bottle on different surfaces, such as carpet, wood floors, tile, etc. *Is it easier to land the bottle upright on some surfaces than others?*

SCIENCE FAIR IDEA

Try the activity with bottles of different sizes or shapes. *Do some work better than others? Do you have a "favorite" type of bottle?*

OBSERVATIONS AND RESULTS

Although results may vary slightly depending on an individual's technique, you probably found you had the most success with a bottle roughly one-quarter to one-third full of water. It was very difficult (maybe almost impossible) to successfully flip either a completely empty or completely full bottle.

The explanation for this phenomenon depends on angular momentum, which you'll remember must be conserved when no outside torque acts on an object and depends on moment of inertia and angular velocity. When a water bottle is spinning through the air, no torque is exerted on it (neglecting air resistance). Also remember the moment of inertia of a rigid object, such as an empty water bottle, does not change as it spins. The empty bottle's angular velocity, therefore, stays the same as it flies through the air just like a spinning coin.

That makes it very hard to control the bottle's descent and difficult to get it to land upright. The same also applies to the completely full bottle. Even though it is full of liquid water, there is no room for the water to slosh around, so the distribution of mass within the bottle remains the same and its angular velocity stays constant.

All that changes when you use a partially filled bottle of water. Initially the water's mass is concentrated at the bottom of the bottle. When you toss the bottle, there is room for the water to slosh around. It spreads out along the bottle's length, increasing the moment of inertia and decreasing the angular velocity (conserving angular momentum). The bottle's spinning slows down as it flies through the air—making it possible, if timed properly, to get the bottle to land upright.

If you try the same trick with ice, even though the bottle is filled the same amount, it doesn't work because the solid ice cannot slosh around.

CLEANUP

Don't forget to recycle your bottle when you are done with it!

Rolling Race

ON YOUR MARK, GET SET, ROLL! WHICH OBJECTS WILL MAKE IT TO THE BOTTOM OF THE RAMP FIRST—AND WHY? GET READY TO SET UP A RAMP AND TEST OUT YOUR THEORIES!

Imagine rolling two identical cans down a slope, but one is empty and the other is full. Which one will reach the bottom first? You might have learned that when dropped straight down, all objects fall at the same rate regardless of how heavy they are (neglecting air resistance). Is the same true for objects rolling down a hill? Try this activity to find out!

PROJECT TIME

20-30 minutes

KEY CONCEPTS

Physics
Mass
Gravity
Kinetic energy
Potential energy

BACKGROUND

When you lift an object up off the ground, it has potential energy due to gravity. The amount of potential energy depends on the object's mass, the strength of gravity, and how high it is off the ground. When you drop the object, this potential energy is converted into kinetic energy, or the energy of motion. Kinetic energy depends on an object's mass and its speed. Ignoring frictional losses, the total amount of energy is conserved.

For a rolling object, kinetic energy is split into two types: translational (motion in a straight line) and rotational (spinning). So when you roll a ball down a ramp, it has the most potential energy when it is at the top, and this potential energy is converted to both translational and rotational kinetic energy as it rolls down. This leads to the question: Will all rolling objects accelerate down the ramp at the same rate, regardless of their mass or diameter?

The answer depends on the objects' moment of inertia, or a measure of how "spread out" their mass is. If two cylinders have the same mass but different diameters, the one with a bigger diameter will have a bigger moment of inertia, because its mass is more spread out. Similarly, if two cylinders have the same mass and diameter, but one is hollow (so all its mass is concentrated around the outer edge), the hollow one will have a bigger moment of inertia. Does moment of inertia affect how fast an object will roll down a ramp? Give this activity a whirl to discover the surprising result!

MATERIALS

- Two soup or bean or soda cans (You will be testing one empty and one full.)
- A hollow sphere (such as an inflatable ball)
- A solid sphere (such as a marble) (It does not need to be the same size as the hollow sphere.)
- Cardboard box or stack of textbooks
- Flat, rigid material to use as a ramp, such as a piece of foam-core poster board or wooden board. The longer the ramp, the easier it will be to see the results.

PREPARATION

- Empty, wash, and dry one of the cans. (Don't waste food—store it in another container!)
- Prop up one end of your ramp on a box or stack of books so it forms about a 10- to 20-degree angle with the floor.

PROCEDURE

- Hold both cans next to each other at the top of the ramp. *Which one do you think will get to the bottom first?*
- Let go of both cans at the same time. Watch the cans closely. *Which one reaches the bottom first?*
- Repeat the race a few more times. *Does the same can win each time?*
- Now try the race with your solid and hollow spheres. *Which one do you predict will get to the bottom first? What happens when you race them?*

SCIENCE FAIR IDEA

Find more round objects (spheres or cylinders) that you can roll down the ramp—for example, rolls of tape, markers, plastic bottles, different types of balls, etc. Try racing different types objects against each other. *What seems to be the best predictor of which object will make it to the bottom of the ramp first?*

SCIENCE FAIR IDEA

Try the activity with cans of different diameters. *What happens if you compare two full (or two empty) cans with different diameters? What about an empty small can versus a full large can or vice versa?*

SCIENCE FAIR IDEA

Try racing different combinations of cylinders and spheres against each other (hollow cylinder versus solid sphere, etc.). *Can you make an accurate prediction of which object will reach the bottom first?*

OBSERVATIONS AND RESULTS

You should find that a solid object will always roll down the ramp faster than a hollow object of the same shape (sphere or cylinder)—regardless of their exact mass or diameter. This might come as a surprising or counterintuitive result! A classic physics textbook version of this problem asks what will happen if you roll two cylinders of the same mass and diameter—one solid and one hollow—down a ramp. The answer is that the solid one will reach the bottom first. In that specific case, it is true that the solid cylinder has a lower moment of inertia than the hollow one does. (Although they have the same mass, all of the hollow cylinder's mass is concentrated around its outer edge, so its moment of inertia is higher.)

However, it is incorrect to say "the object with a lower moment of inertia will always roll down the ramp faster." It takes a bit of algebra to prove, but it turns out that the absolute mass and diameter of the cylinder do not matter when calculating how fast it will move down the ramp—only whether it is hollow or solid. So, in this activity you should have found that a full can of beans rolls down the ramp faster than an empty can—even though it has a higher moment of inertia. (It has the same diameter, but is much heavier than an empty aluminum can.) Applying the same concept shows that two cans of different diameters should roll down the ramp at the same speed, as long as they are both either empty or full. The same principles apply to spheres as well—a solid sphere, such as a marble, should roll faster than a hollow sphere, such as an air-filled ball, regardless of their respective diameters.

CLEANUP

Recycle the empty cans and put away the other items you used.

THE SCIENTIFIC METHOD

The scientific method helps scientists—and students—gather facts to prove whether an idea is true. Using this method, scientists come up with ideas and then test those ideas by observing facts and drawing conclusions. You can use the scientific method to develop and test your own ideas!

Question: What do you want to learn? What problem needs to be solved? Be as specific as possible.

Research: Learn more about your topic and refine your question.

Hypothesis: Form an educated guess about what you think will answer your question. This allows you to make a prediction you can test.

Experiment: Create a test to learn if your hypothesis is correct. Limit the number of variables, or elements of the experiment that could change.

Analysis: Record your observations about the progress and results of your experiment. Then analyze your data to understand what it means.

Conclusion: Review all your data. Did the results of the experiment match the prediction? If so, your hypothesis was correct. If not, your hypothesis may need to be changed.

GLOSSARY

adjacent: Next to or near something.
capillary: Having a long, slender form and a small inner diameter; involving, held by, or resulting from surface tension.
comprise: To be made up of.
counteract: To make something have less of an effect or no effect at all.
diameter: A straight line that runs from one side of a figure and passes through the center.
dimension: One of three or four points determining a position in space or space and time.
exploit: To get value or use out of something.
granular: Made up of small grains.
humidifier: A machine that creates humidity, or moisture in the air.
intuitive: Known by feeling rather than thought; instinctive.
neglect: To ignore.
nozzle: A short tube often used on the end of a hose or pipe to direct or speed up a flow of fluid.
optimal: The most desirable or satisfactory.
parameter: A measurement or value on which something else depends.
perpendicular: Standing at a right angle.
phenomenon: An observable fact or event.
physics: A science that deals with the facts about matter and motion and includes the subjects of mechanics, heat, light, electricity, sound, and the atomic nucleus.
repose: A state of resting.
simulated: Something fake that is made to look real.
technique: A way of doing something using special knowledge or skill.
theoretical: Existing only as an idea.

ADDITIONAL RESOURCES

Books

Hardy, Freya, and Sara Mulvanny. *The Book of Big Science Ideas*. Brighton, UK: Ivy Kids, 2019.

Minter, Laura, and Tia Williams. *Science School: 30 Awesome Science Experiments to Try at Home*. East Sussex, UK: Button Books, 2021.

Tolish, Alexander. *Gravity Explained*. New York, NY: Enslow Publishing, 2019.

Websites

Discovery Education
sciencefaircentral.com

Exploratorium
https://www.exploratorium.edu/search/science%20fair%20projects

Science Buddies
https://www.sciencebuddies.org/science-fair-projects/project-ideas/list

Science Fun
https://www.sciencefun.org/?s=science+fair

Videos

"Gravity and the Expanding Universe"
https://ny.pbslearningmedia.org/resource/phy03.sci.ess.eiu.expand/gravity-and-the-expanding-universe/, PBS Learning Media, 4:03.

"GPS: Gravity Fountain"
https://ny.pbslearningmedia.org/resource/39bc5591-5ed8-4fa2-81b9-f9cfa50d76c7/39bc5591-5ed8-4fa2-81b9-f9cfa50d76c7/, PBS Learning Media, 7:03.

INDEX